SpringerBriefs in Molecular Science

Protein Folding and Structure

Series editor

Cláudio M. Gomes, Biosystems & Integrative Sciences Institute,
Faculdade de Ciências da Universidade de Lisboa, Lisboa, Portugal

About the Series

Prepared by leading experts, the series contains diverse types of contributions, from snapshot volumes that allow fast entry to a general topic to those covering more specialized aspects in the field of protein folding and structure. In common, these *Briefs* aim at covering essential concepts, methodologies and ideas in the context of contemporary research in protein science. Through these compact volumes, this series serves as a venue for publication between typical research papers, review articles and full books, and aims at a broad audience, from students to researchers in academia and industry.

About the Editor

Cláudio M. Gomes joined the Gulbenkian Ph.D. program in Biology and Medicine (1994) and obtained his Ph.D. (1999) and habilitation (2013) in Biochemistry from the Universidade Nova de Lisboa. He is currently Coordinating Group Leader Researcher and a consolidation-level fellow of the FCT Investigator Programme, at the Biosystems and Integrative Sciences Institute at the Faculty of Sciences University of Lisboa, after over a decade as group leader and assistant researcher at Universidade Nova de Lisboa (2003–2014). His research is focused on protein folding, misfolding and aggregation, mostly concerning molecular mechanisms in neurodegenerative and metabolic diseases. He is involved in extensive publishing and editorial activities, as an author in scientific journals, member of the Editorial boards of several scientific journals and editor of thematic journals issues and books.

More information about this series at http://www.springer.com/series/11958

Jenny Presto · Jan Johansson

The BRICHOS Domain

Its Proproteins and Functions

 Springer

Jenny Presto
Department of Neurobiology,
  Care Sciences and Society,
  Center for Alzheimer Research
Karolinska Institutet
Huddinge
Sweden

Jan Johansson
Department of Neurobiology,
  Care Sciences and Society,
  Center for Alzheimer Research
Karolinska Institutet
Huddinge
Sweden

and

Department of Anatomy, Physiology
  and Biochemistry
Swedish University of Agricultural
  Sciences
Uppsala
Sweden

ISSN 2191-5407   ISSN 2191-5415   (electronic)
SpringerBriefs in Molecular Science
ISSN 2199-3157   ISSN 2199-3165   (electronic)
Protein Folding and Structure
ISBN 978-3-319-16563-9   ISBN 978-3-319-16564-6   (eBook)
DOI 10.1007/978-3-319-16564-6

Library of Congress Control Number: 2015934046

Springer Cham Heidelberg New York Dordrecht London

Springer International Publishing AG Switzerland is part of Springer Science+Business Media (www.springer.com)

Preface

In the second volume of the *SpringerBriefs series* on *Protein Folding and Structure* we now turn to protein domains. Protein domains are compact structural units with well-defined tertiary interactions, whose folding and evolution is frequently independent of those of the proteins in which they are found. Classical protein domains include Zinc fingers, EF-hands and Immunoglobulin-like domains to which precise functions are associated. However, the fact is that a substantial number of proteins comprise protein domains with as yet uncharacterised functions. Understanding proteins and their structural and functional versatilities thus requires a thorough knowledge of protein domains and potential functions thereof. In this volume, Presto and Johansson present us with an insightful assay which introduces the BRICHOS domain, a structural module which renders chaperone and anti-aggregation properties in its proproteins. The volume uncovers how knowledge on BRICHOS has evolved since its discovery over a decade ago, overviews the different proteins in which it is found, and accounts for the latest discoveries relating to new regulatory functions of this domain over amyloid-β aggregation, with implications in Alzheimer's Disease. By combining a historical perspective with the latest biochemical and functional insights into BRICHOS domains, this volume constitutes an excellent addition to the series that illustrates the relevance of frontier research on protein domains towards the goal of understanding protein structure and folding in a broader sense. Enjoy reading.

Lisboa, March 2015
Cláudio M. Gomes
Editor, SpringerBriefs Series on
Protein Folding and Structure

Acknowledgments

We would like to thank all the current and former members working in our group with the BRICHOS domain, as well as our collaborators, who have contributed to increasing the knowledge about the BRICHOS domain and its functions. This work was supported by the Swedish Research Council, The Swedish Alzheimer foundation, The Åke Wibergs foundation, The Magn Bergvalls foundation, Foundation of Gamla tjänarinnor, and The Loo and Hans Ostermans foundation for geriatric research.

Contents

Abbreviations

AD	Alzheimer's disease
APP	Amyloid(-β) precursor protein
Aβ	Amyloid β-peptide
BACE	Beta-site APP cleaving enzyme
BBB	Blood brain barrier
ER	Endoplasmic reticulum
FBD	Familial British dementia
FDD	Familial Danish dementia
ILD	Interstitial lung disease
NFT	Neurofibrillary tangles
RDS	Respiratory distress syndrome
SP-B	Surfactant protein B
SP-C	Surfactant protein C
ThT	Thioflavin T
TM	Transmembrane

About the Authors

Jenny Presto obtained her Ph.D. in Medical Biochemistry from Uppsala University in 2006, where she studied the biosynthesis of heparan sulfate. She did a postdoc at the Swedish University of Agricultural Sciences (SLU) in Uppsala, Sweden, studying the function of the BRICHOS domain as a chaperone for surfactant protein C.

In 2011, Jenny moved to Karolinska Institutet, Center for Alzheimer Research, Stockholm, Sweden, where she currently is an assistant professor. Her research focuses on the *in vitro* and *in vivo* inhibition of amyloid-β fibril formation and toxicity by the chaperone BRICHOS.

Jan Johansson obtained his M.D. from Karolinska Institutet in 1990, and got a Ph.D. from Karolinska Institutet 1991. He spent a postdoc with Kurt Wüthrich at the ETH in Zürich during 1992 and 1993, which resulted on the first structure determination of a lung surfactant protein. He is currently professor of Biological Dementia Research at Karolinska Institutet, and of Medical Biochemistry at the Swedish University of Agricultural Sciences (SLU), Uppsala, Sweden.

Johansson's research has generated synthetic proteins for treatment of respiratory distress and has led to identification of fundamental properties of amyloidogenic proteins, defined a novel type of chaperone defense involved in amyloid diseases, and revealed pathways to control protein assembly in spider silk formation.

Abstract

The BRICHOS domain is present in several protein families associated with respiratory distress, amyloid disease, dementia and cancer. The proproteins are known or predicted to be type II transmembrane or secretory proteins, with the BRICHOS domain facing the endoplasmic reticulum lumen. When it was first described, it was proposed to have intramolecular chaperone-like function, and it has later been suggested that BRICHOS domains bind to aggregation-prone regions in their respective precursor protein, and thereby protect them from misfolding. This has been shown for the proSP-C BRICHOS domain, in which mutations lead to lung disease and amyloid formation of the extremely β-prone segment in the transmembrane region of proSP-C. This endogenous anti-amyloid activity seems to extend to other amyloidogenic peptides, and the BRICHOS domains of Bri2, proSP-C, and Gastrokine-1 have so far proven to interact with and inhibit aggregation and fibril formation of the Alzheimer's disease associated peptide Aβ. The mechanism by which proSP-C BRICHOS inhibits Aβ fibril formation was recently revealed and it was found that BRICHOS specifically blocks the secondary nucleation pathway. The secondary nucleation pathway is believed to be the main source of toxic oligomer formation during Aβ fibril formation, which makes the proSP-C BRICHOS the first-known chaperone domain that specifically targets and decreases the toxicity associated with the formation of amyloid fibrils. Recent results from a *Drosophila melanogaster* model show that BRICHOS prevents the *in vivo* toxicity associated with Aβ aggregation, which suggests that targeting this chaperone can be fruitful as a treatment for AD.

Keywords Amyloid · Chaperone · Alzheimer's disease · Protein folding · Protein aggregation disease

The BRICHOS Domain

Its Proproteins and Functions

1 Introduction to BRICHOS Superfamily

A diverse group of 8 protein families was in 2002 found to contain a domain—
BRICHOS—that has low sequence conservation in between the different families
[1], but shares a remarkable similarity in predicted secondary structure [2]. The
name BRICHOS was given from combining the names of three proteins in the
superfamily, *Bri*2, *Cho*ndromodulin-1 and *S*urfactant protein C (SP-C), and several more families have been added to the superfamily since BRICHOS first was
described (Fig. 1a) [1, 2]. Today more than 1000 proteins containing a BRICHOS
can be found by bioinformatic tools (see http://smart.embl-heidelberg.de/smart/
do_annotation.pl?DOMAIN=BRICHOS) [1]. The approximately 100 amino acid
residue large BRICHOS domain is present in proteins found in species from prim-
itive marine organisms to humans, and they all share a common architecture and
are known or predicted to be type II transmembrane (TM) or secretory proteins,
with the BRICHOS domain facing the endoplasmic reticulum (ER) lumen [1]. The
BRICHOS containing proteins moreover contain a short N-terminal cytosolic segment with largely unknown functions (but it has been shown to be important for
correct intracellular targeting of proSP-C [3]), and a linker segment in between
the TM part and the BRICHOS domain. There are only three strictly conserved
amino acid residues in the BRICHOS domain, one Asp and two Cys, where the
two cysteines likely form a disulphide bridge (Fig. 1b) [1, 2].

For many of the BRICHOS protein families the functions remain to be
explored, but the BRICHOS domain has so far been linked to diseases like
familial dementia (Bri2 and Bri3), Alzheimer's disease (Bri2) respiratory disease (proSP-C), amyloidoses (Bri2 and proSP-C), and cancer/tumor suppression (chondromodulin-I and gastrokine 1 and 2) (Table 1) [1, 4–6]. One common
attribute of the BRICHOS containing proteins is that they all (with one exception) contain a C-terminal region with high β-sheet propensity, predicted to
form a strand-loop-strand β-hairpin conformation [2]. The exception refers

© The Author(s) 2015
J. Presto and J. Johansson, *The BRICHOS Domain*, Protein Folding and Structure,
DOI 10.1007/978-3-319-16564-6_1

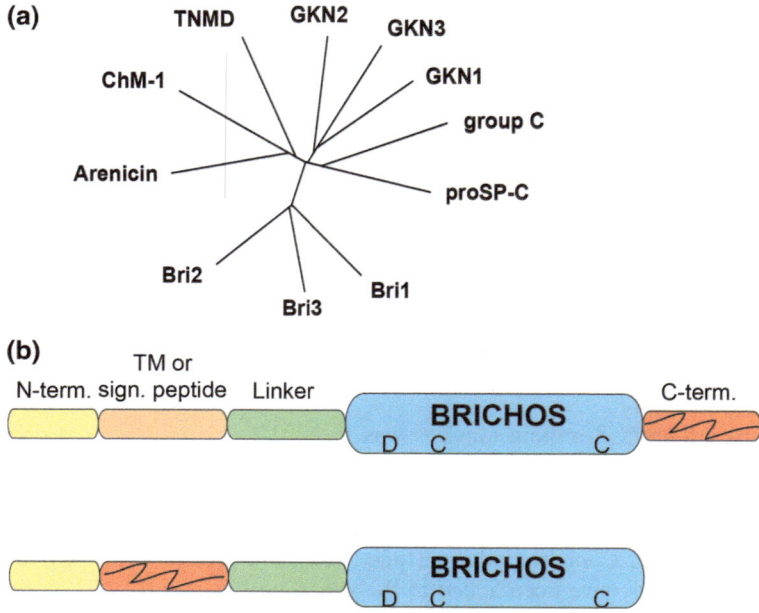

Fig. 1 BRICHOS protein families and their overall architecture. Evolutionary tree of the different BRICHOS domain-containing proteins. Distances reflect their evolutionary separation (**a**). BRICHOS containing proteins have a N-terminal cytosolic segment, a TM or a signal peptide region, a linker segment followed by the BRICHOS domain. A C-terminal domain with high β-sheet propensity is present in all BRICHOS containing proteins (*upper graph*), except for proSP-C that shows extremely high β-sheet propensity in its TM region (*lower graph*). The aggregation prone regions are marked with *wavy black lines* (**b**). **a** was originally published in [61]

to proSP-C, which lacks a C-terminal β-hairpin but instead has a TM region with extremely high β-sheet propensity, which is proposed to be able to form a β-hairpin structure together with the linker domain of proSP-C [7–9]. The BRICHOS domain of proSP-C was the first BRICHOS to be experimentally investigated [10], which resulted in a proposed physiological function of this BRICHOS domain; to work as a chaperone preventing misfolding and amyloid formation of the TM region of proSP-C [9]. This chaperone action of proSP-C BRICHOS was hypothesized to be intramolecular, but an intermolecular action of BRICHOS cannot be ruled out. In fact, it has been shown that expression of the proSP-C BRICHOS domain in trans rescues misfolding and aggregation of mutant proSP-C in a human cell line [11]. The discovery of BRICHOS "anti-amyloid" activity describes a scientific journey, where a lung specific protein has given new hope for finding a treatment strategy for Alzheimer's disease (AD)

Table 1 BRICHOS proteins, which have been found to be associated with human diseases

BRICHOS containing protein	Disease association	Role of BRICHOS protein in the disease
Bri2	Familial British and Danish Dementia (FBD and FDD) (amyloid diseases)	Mutations in Bri2 lead to amyloidogenic peptides ABri and ADan [87]
proSP-C	Interstitial lung disease (amyloid disease)	Mutations in the BRICHOS domain or the linker region lead to amyloid formation of SP-C [9]
Gastrokine-1	Gastric cancer	GKN1 expression is downregulated in gastric adenocarcinoma tissues [66, 67]
Gastrokine-2	Gastric cancer	GKN2 expression is downregulated in gastric adenocarcinoma tissues [68]
Chondromodulin-1	Chondrosarcoma	Decreased expression of CHM-1 in chondrosarcoma [4]
Bri2	Alzheimer's disease	Changes in their gene expression have been correlated with late-onset AD and ApoE4 carriers [100]

and other amyloid diseases, by harnessing the BRICHOS domain as one of nature's amyloid-specific chaperones.

2 BRICHOS in proSP-C

2.1 Lung Surfactant and SP-C

Insufficient amounts of lung surfactant result in respiratory distress syndrome (RDS) and often occur in premature children with a gestational age below about 30 weeks. RDS can nowadays be treated via instillation of surfactant preparations obtained from animal (mainly pig or cow) lungs and this therapy is very effective [12]. However, synthetic surfactant preparations with equal or better effects than the naturally derived counterparts have been searched for since the 1980s, when molecular studies of surfactant proteins gained momentum [13]. A fully synthetic surfactant would avoid potential problems like spreading of infectious agents and immunogenicity associated with surfactant extracted from animal lungs, and may be less expensive to manufacture. A surfactant preparation based on a designed model peptide is available in the United States [14], but no synthetic surfactant that contains proteins and that mimic the natural derived products has yet reached the market [15].

Initially, natural derived lung surfactant preparations were considered to be protein-free, probably because it was found unlikely that organic solvent extracts from a very lipid-rich source could contain proteins. When their existence and importance for surfactant function was acknowledged several different approaches were tried to obtain pure surfactant protein preparations. A successful method introduced size exclusion chromatography in acidified chloroform/methanol mixtures [16], an unusual method for obtaining native proteins. By this approach, SP-C and surfactant protein B (SP-B) could eventually be isolated and thereafter structurally characterized [17, 18]. SP-C was first purified to homogeneity from porcine lung tissue and later also from other sources, like bronchoalveolar lavage and amniotic fluids, and from different animal species [19]. The ensuing studies of SP-C from various angles revealed that it is a unique protein from several aspects—biochemical and cell biological properties, sites of expression and structure [20]. SP-C is exclusively produced in the alveolar type II cell and it is the only known example of an extracellular TM protein, and its structure is unique as well, see further below. SP-C behaves as a lipid from solubility point-of-view, being insoluble in aqueous solutions and requiring organic solvents to stay in solution in the absence of a lipid environment. This results in that SP-C elutes with the phospholipid fraction when an organic extract of lungs is separated over a Lipidex 5000 column in organic solvents [21]. This purification step was used to isolate a natural derived surfactant preparation for treatment of RDS in premature babies and as a consequence, the presence of SP-C in the phospholipid fraction was not realized until at later stage when attempts were made to remove protein contaminants. The solubility properties of SP-C reflects its localization, SP-C is inserted as a TM peptide in phospholipid bilayers and only a minor part of it interacts with the polar phospholipid head-groups and aqueous surroundings [22].

The ability to isolate SP-C made it possible to characterize its primary, secondary and tertiary structures in organic solvents as well as in phospholipid and detergent matrices. This revealed that SP-C is a 35-residue covalent lipopeptide—it contains two (or one in minks and dogs) palmitoyl chains linked via thioesters to Cys in the N-terminal region and its structure is composed of an α-helix spanning residues 9–34 [23, 24]. This α-helix is perfectly sized to span a bilayer of dipalmitoylphosphatidylcholine in a TM orientation, and when inserted into a phospholipid monolayer, which is supposed to exist at the air-water interface in the alveoli, SP-C is tilted in order to maximize the interactions between its mainly unpolar α-helix and the lipid acyl chains (Fig. 2a) [25, 26]. Although SP-C has been studied in great detail from several aspects, see below, its molecular actions in the lung surfactant system are still not clarified. Based on its unique molecular properties and site of expression, it is generally assumed that SP-C performs specific functions in the surfactant system. SP-C accelerates the spreading of phospholipids at an air-water interface and it affects the packing of phospholipid bilayers, but how this translates into a specific function that relates to the lowering of alveolar surface tension by lung surfactant is mainly unknown [27].

2.2 The SP-C Structure and Its Unique Properties

The SP-C α-helix is made up of a fourteen-residue stretch covering residues 15–28 that, except for a strictly conserved Leu at position 22, contains

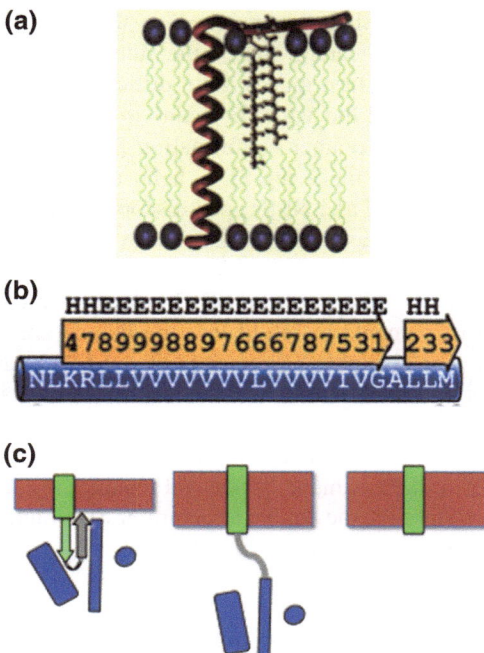

Fig. 2 The mature SP-C is composed of an α-helix that perfectly fits in a transmembraneous orientation into a bilayer composed of dipalmitoylphosphatidylcholine, the main component of lung surfactant, see Ref. [21] for details. The N-terminal eight residues of SP-C are structurally disordered and a thio-ester palmitoyl residue covalently modifies the Cys residues at positions five and six [22] (**a**). The many consecutive Val residues in SP-C give rise to that the α-helix determined by NMR (depicted as a *blue cylinder* and the amino acid sequence of the helical part of porcine SP-C is given) has a very high β-sheet propensity, as predicted by the neural network prediction program. Protein profiling of Heidelberg (PHD) (depicted as a *yellow arrow* with reliability indices plotted, where nine is highest possible reliability) or by the Chou-Fasman secondary structure propensities (*H* stands for helix and *E* stands for extended, i.e. β-strand conformation). An α-helix that is composed of residues that are strongly predicted to form a β-strand is called a discordant helix [7] (**b**). Proposed mechanism of action of the proSP-C BRICHOS domain (*blue*) in the ER (the ER membrane is shown in *brown*), where it assists the proSP-C TM region (*green*) to form the mature α-helical structure with the help of a part of the linker region (*gray*). This function of the BRICHOS domain is crucial for the membrane insertion of the SP-C part. After proper TM insertion of the SP-C part, the linker and BRICHOS domain are cleaved off by proteases, thus generating mature SP-C that eventually is secreted together with phospholipids into the alveolar space. The fate or possible functions of the BRICHOS domain after release from proSP-C are not known [9] (**c**). **a** was originally published in [12]. **b** was originally published in [7] © the American Society for Biochemistry and Molecular Biology. **c** was originally published in Proc Natl Acad Sci USA, Feb 14, 2012; 109(7):2325–2329 [9]

exclusively unpolar residues that are branched on the β-carbon, i.e. Val and Ile [18]. The branching on the β-carbon makes the side-chains bulky close the polypeptide backbone, which in turn makes them difficult to pack into an α-helical structure. In line with this all the Val side-chains in the SP-C poly-Val helix are locked in one and the same rotamer conformation [24]. This locking is entropically unfavourable and could make the SP-C helix energetically unstable. Using hydrogen-deuterium exchange and NMR spectroscopy it was shown that the SP-C poly-Val helix is indeed metastable (i.e. kinetically but not thermodynamically stable) and that the lifetime of the helical structure equals the lifetime of soluble peptide [8]. As a result of this, after the SP-C helix has unfolded into an unordered conformation it is unable to refold into the helical conformation before it goes out of solution. The insoluble aggregated form of SP-C turned out to contain mainly β-sheet structure and showed abundant amyloid-like fibrils [28]. The native α-helical state of SP-C will thus only be maintained as long as the energetic barrier to unfolding is not overcome. This barrier was estimated to about 100 kJ/mol, which is higher than for helices made of non-Val residues [8]. The presence of the two palmitoyl groups stabilize the helical structure of SP-C and reduces the rate of unfolding and amyloid fibril formation [29, 30].

2.3 Solving the SP-C Folding Problem by Rational Design

When the primary structure of SP-C had been established, synthesis of SP-C or analogues with similar sequences was considered to be an attractive route to design of a fully synthetic surfactant for replacement therapy of RDS. Early attempts to produce α-helical SP-C peptides via organic synthesis failed [31, 32]. In retrospect this can be explained by the observed metastability of helical SP-C versus non-helical and β-sheet SP-C peptides [8]. The high β-sheet propensity of Val results in that synthetic replicas mainly form β-sheet aggregates, which have very low activity as components in surfactant preparations (Fig. 2b).

The side-chain of Leu has one methylene group more than Val, which places the branch of the side-chain further away from the polypeptide backbone and it can thus rotate also when it is packed in an α-helical conformation. As a consequence of this, Leu has a high helix propensity [33]. Poly-Val substitution for poly-Leu, in a synthetic analogue of SP-C results in dramatic increase in spontaneous formation of a stable α-helix [7, 33]. By replacing all Val in SP-C with Leu a synthetic peptide that spontaneously folds into an α-helical conformation and is active as a surfactant when combined with synthetic phospholipids was obtained [34]. This SP-C(Leu) peptide, in sharp contrast to native SP-C, was found not to unfold into β-sheet aggregates and stays in solution for weeks [7]. A fully synthetic surfactant based on SP-C(Leu) is equally or more efficient than the porcine-derived surfactant Curosurf (the today most sold surfactant preparation) in animal models of RDS or surfactant inactivation [35, 36], and is currently in clinical trials for treatment of RDS in premature children.

2.4 Solving the SP-C Folding Problem by the BRICHOS Domain—How Nature Solved It

ProSP-C is expressed by only one cell type in the body, the alveolar type II epithelial cell of the lung [37, 38]. Human proSP-C is a 197 amino acid integral membrane proprotein that is processed to the 35 amino acid mature peptide within the secretory pathway of the type II cell (Fig. 3) [39, 40]. The mature peptide (residues 24–58 of proSP-C) encodes a signal sequence that directs the newly synthesized proprotein into the ER membrane [3, 41]. Two thirds of the 35 amino acid mature peptide reside in the TM domain while the remaining N-terminal portion is located in the cytosol. The latter domain is palmitoylated at adjacent cysteine residues and is flanked by an N-terminal propeptide (residues 1–23). Trafficking of proSP-C from the ER to the distal secretory pathway is dependent on the cytosolic N-terminal propeptide [3, 42]. In contrast, the lumenal C-terminal domain of proSP-C is dispensable for intracellular trafficking and secretion of SP-C, once it has formed a TM α-helix. Trafficking of proSP-C to the multivesicular body in the distal secretory pathway is necessary for processing to the mature peptide [3, 39]. Proteases in the lumen of the multivesicular body remove the C-terminal peptide from proSP-C in at least two steps. Cathepsin H has been implicated in this process [43], but the precise

Fig. 3 BRICHOS domain in proSP-C. The BRICHOS domain is located in the C-terminal part of proSP-C. BRICHOS works as a chaperone for the valine-rich TM part of proSP-C, which after processing into mature SP-C in the secretory pathway of alveolar type II cells is stored and secreted in lamellar bodies, as a constituent of lung surfactant. The aggregation prone region in the TM segment is marked with *wavy black lines*

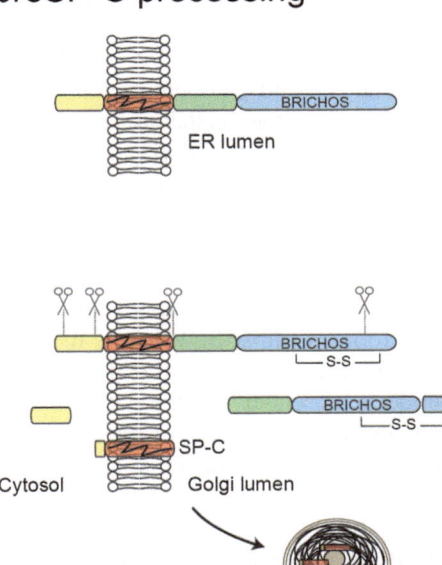

proSP-C processing

number and identity of proteases involved in maturation of proSP-C is not known [3, 44]. Fusion of the multivesicular body with a lamellar body (a secretory granule specialized for intracellular storage of surfactant) leads to incorporation of the SP-C-containing vesicles into the highly packed bilayer membranes of the lamellar body. The lipid bilayer contents of the lamellar body, including SP-B and SP-C, are released into the alveolar airspace where they unravel and ultimately contribute to formation and maintenance of a phospholipid rich film (pulmonary surfactant) along the surface of the epithelium.

The inability of the SP-C α-helix to refold in vitro raises questions about its formation in vivo. It has been shown that a model poly-Val segment is less efficient in forming an α-helix in the ribosome and ER translocon than a poly-Leu segment, which folds into an α-helix already in the ribosome [45]. We have discovered that a BRICHOS domain of proSP-C works as a molecular chaperone that prevents aggregation of the TM SP-C part, and promotes its folding into a helix (Fig. 2c) [9, 10, 46–49]. Intriguingly, mutations that segregate with human lung fibrosis, or interstitial lung disease (ILD), are localized to the BRICHOS domain, and our recent determination of the first crystal structure of a BRICHOS domain can explain how such mutations inactivate its chaperone function [9]. An important experimental support for the notion that the function of the proSP-C BRICHOS domain is to prevent misfolding of the proSP-C poly-Val TM segment came from the observation that expression of proSP-C with a poly-Leu TM segment (thus mimicking the replacements made to allow synthesis of thermodynamically stable SP-C analogue for artificial surfactant, see above) instead of the native poly-Val counterpart in a cell line results in strong reduction of aggregation even in the presence of a non-functional BRICHOS domain [11].

Available data suggest that proSP-C BRICHOS acts as a chaperone for the extremely hydrophobic and β-structure-prone TM proSP-C segment during biosynthesis [49]. Based on the crystal structure of the proSP-C BRICHOS domain (Fig. 4a) a model for its function during proSP-C biosynthesis was suggested [9]. According to this model the BRICHOS domain, in a monomeric state, together with the linker region captures parts of the proSP-C TM region, which have not folded into an α-helix and been properly inserted into the ER membrane. Once the proSP-C TM region has been properly membrane-inserted, the subsequently exposed linker region becomes proteolytically cleaved and the BRICHOS domain is released and forms a trimer in which its hydrophobic client-binding surface is buried. The proposal that the BRICHOS domain with the help of the linker region captures peptides representative of the poly-Val TM part of proSP-C, may explain how mutations in the linker region or the BRICHOS domain both can be associated with ILD and amyloid formation. The linker region may serve as a substitute β-strand, which docks to the BRICHOS-bound proSP-C TM region, forming a β-hairpin structure (Fig. 4b). Such a function of the linker would explain both why proSP-C is the only BRICHOS-containing protein with a highly conserved linker region and the only one with a presumed target region composed of a single β-strand. Unpublished data show that the folding of the SP-C TM part during proSP-C translation indeed is inefficient. Furthermore, the proSP-C BRICHOS

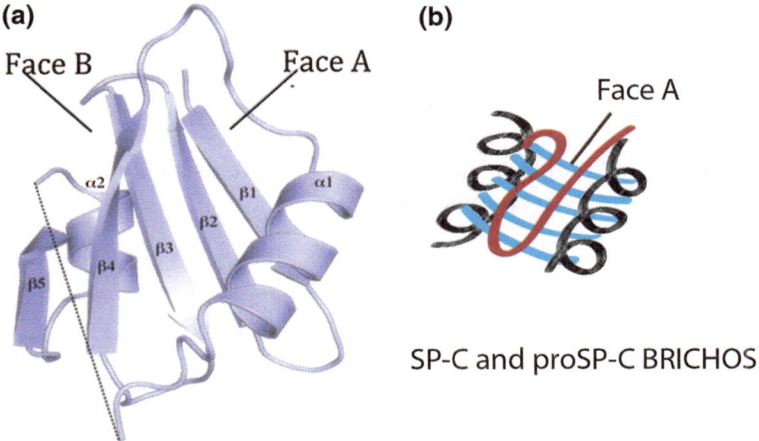

(a)

Face B   Face A

α2   β1   α1
β2
β3
β4   β5

(b)

Face A

SP-C and proSP-C BRICHOS

Fig. 4 Structure of proSP-C BRICHOS. The crystal structure of the proSP-C BRICHOS domain is built up of two α-helices, one on each side of a central five-stranded β-sheet. The β-sheet side called face A is believed to be the binding site for target peptides, and the *dashed line* indicates the missing region in the crystal structure between helices α1 and α2 [9] (**a**). Face A of the BRICHOS domain (*blue*) of proSP-C binds to non-helical hydrophobic peptides (*red*), and is proposed to bind to a β-hairpin motif. The two helices found in the crystal structure of proSP-C BRICHOS are coloured *black* [9] (**b**). a was originally published in Proc Natl Acad Sci USA, Feb 14, 2012; 109(7):2325–2329 [9]

can insert into an ER-like membrane and the insertion is promoted by unfolded poly-Val peptides present in the membrane, which supports the theory that the domain is capable of capturing the SP-C TM part by dipping into the membrane, and fulfil its chaperone activity (Sáenz A., Presto J., Lara P., Akinyl Oloo L., García-Fojeda B., Nilsson I.M., Johansson J., and Casals C., 2015, "Folding and intramembraneous BRICHOS binding of the proSP-C transmembrane segment", submitted).

Knock-out of the surfactant protein C gene (*SFTPC*) in mice results in neonatally viable mice [50]. This is in contrast to the situation for SP-B, for which genetic ablation in mice [51] and mutations in humans that result in lack of functional SP-B [52], result in neonatal lethality [53]. The proSP-C knockout mice however show some deficits of surfactant function in vitro [50] and develop diffuse lung disease over time [54]. The pathology of ILD associated with mutations in human SFTPC, in particular the part coding for the BRICHOS domain, is not recapitulated in SFTPC knock-out mice. Lack of detectable mature SP-C in alveolar surfactant has been described in ILD patients with SFTPC mutation [55], but the fact that the pathology seen in proSP-C knockout mice differs from that in ILD patients with BRICHOS mutations argue for that the latter is a result of gain of a toxic function rather than lack of mature SP-C. However, it should be pointed out that the ILD pathology is complex and also varies significantly between patients, so further research is required to fully understand the pathophysiology associated with mutations in proSP-C.

We have found that proSP-C BRICHOS mutations can lead to amyloid disease; BRICHOS is thus the first described example of an endogenous factor that guards extraordinarily amyloidogenic peptides [49]. Such factors are attractive candidates to harness for treatment of amyloid diseases in general, and our recent findings that the BRICHOS domain prevents amyloid β-peptide (Aβ) fibril formation in vitro and in vivo [56–58] have motivated us to explore its ability to treat and prevent Alzheimer's disease, see further below.

2.5 BRICHOS Domain Structure

The first crystal structure of a BRICHOS domain, from proSP-C, was published in 2012. The BRICHOS monomer is built up by two α-helices, one on each side of a central five-stranded β-sheet (Fig. 4) [9]. In the crystal structure, the BRICHOS domain forms a trimer, which also is supported by size-exclusion chromatography of the recombinant protein [59], and trimer formation is most likely a capping mechanism that protects the hydrophic surfaces of the domain. Peptide binding experiments suggests that it is the monomeric BRICHOS that is the active form [60], and one of the sides of the central β-sheet, named face A, is proposed to be the binding site for target peptides [9]. A recent homology modeling of BRICHOS domains from different families suggests that the binding face A is complementary to their respective proposed substrate peptides, that is the C-terminal segments with high β-sheet propensity in the case of all BRICHOS containing proteins except proSP-C, which lacks this region but instead has a β-prone TM segment. This observation supports the hypothesis that these peptides indeed are the physiological targets for the respective BRICHOS domain [61].

3 Proteins with a BRICHOS Domain

The name BRICHOS is derived from the three proteins Bri2 (a member of the Bri family also referred to as integral transmembrane proteins), chondromodulin-I (also known as leukocyte cell-derived chemotaxin), and SP-C [1]. Other BRICHOS-containing proteins encompass gastrokines [62], tenomodulins (also known as chondromodulin-like proteins), arenicins (antimicrobial proteins from lugworms), and the group C proteins (also referred to as C16Orf79, human chromosome 16 open reading frame 79). The BRICHOS domain is the only region of the BRICHOS-containing proteins that is conserved across the entire superfamily [2].

Hedlund et al. [2] searched the UniProtKB and found a total of 139 proteins having a BRICHOS domain, including some that were grouped in undefined groups denoted A, B and C. Today more than 1000 proteins have been found to contain a BRICHOS domain (http://smart.embl-heidelberg.de/smart/do_annotation.pl?DOMAIN=BRICHOS). Gastrokines 1 and 2 and group B are closely related families that are also co-localised in the genome. Group B is today referred to as gastrokine 3. While group B has so far only been found in mouse, rat, cow and dolphin, gastrokines are found in a wide range of mammals. Chondromodulins and tenomodulins are widespread in vertebrates ranging from fish through armadillo and elephant to human. The chondromodulins and tenomodulins genes are located on different chromosomes in humans.

As described above, the proSP-C BRICHOS domain has been suggested to interact with the proSP-C TM region during SP-C biosynthesis. The proSP-C TM region is strongly predicted to form β-sheet, but the same phenomenon is not found in any of the other BRICHOS containing proteins, i.e. their TM regions are, like the clear majority of TM regions, predicted to form α-helices. If one assumes that the apparently conserved BRICHOS structure has a similar function in all proteins, this raises the question: what do the non-proSP-C BRICHOS domains interact with? It was observed that the regions located C-terminally of the BRICHOS domain in all other BRICHOS containing proteins than proSP-C, contain regions that are predicted to form β-hairpin structures [2]. In the case of arenicin, NMR data show that the C-terminal region indeed forms a β-hairpin in solution [63]. The trefoil factor that binds to the gastrokine 2 BRICHOS domain likewise forms a hairpin structure in solution [64], and gastrokine 1 BRICHOS from chicken gizzard binds to a strand-turn-strand structure motif in filamentous actin [65]. These experimental results support the supposition that BRICHOS binds to β-hairpin structures.

3.1 Gastrokines

The gastrokines contain a BRICHOS domain, and their expression levels have been associated with gastric cancer. Gastrokine 1 is expressed in gastric mucosa, but not in gastric carcinoma cells and has a suggested role in mucosal protection [66, 67]. Gastrokine 2, also called TFIZ1 and blottin (the mouse homologue), has been shown to interact with members of the trefoil factor family (TFF1 and TFF2), which are associated with cancer. Gastrokine 2 expression is furthermore decreased in gastric cancer tissue and up-regulation has been shown to suppress tumourigenic and metastatic capacity [68]. The human TFF1 forms a heterodimer with gastrokine 2 [64], and blottin has been found to bind to TFF2 [69]. A novel gastrokine, gastrokine 3, was recently found to be associated with gastric atrophy in mouse and possibly in the host response to *Helicobacter pylori*. However, humans seem to have lost the gastrokine 3 gene expression [5].

3.2 Chondromodulin and Tenomodulin

The chondromodulin-I, also called leukocyte cell derived chemotaxin 1 was one of the first BRICHOS containing proteins found [1]. The 335 amino acid residue precursor protein is post-translationally processed intracellularly into a 28 kDa glycosylated protein that includes the BRICHOS domain, and is found in lysates from cultured chondrocytes and secreted into the surrounding medium. The chondromodulin-I is processed by furin, which cleaves off the C-terminal segment of the protein, like for the processing of Bri2 [70]. Chondromodulin-I is associated with chondrosarcoma and loss of its expression seems to be important for angiogenic properties related to malignant transformation [4]. Tenomodulin is expressed in eye, skeletal muscle, whole rib and dense connective tissue, and its expression is increased during mouse embryonic development [71, 72].

3.3 Group C

The members of the group C family have only been studied at the gene or transcript levels, and the functions of these proteins are unknown. It is, however, notable that the sequence of the C-terminal region of group C, including four Cys, is strictly conserved, suggesting that it is functionally important [2]. Moreover, the Cys-containing regions contain two, strongly predicted β-strands. This suggest that the C-terminal region of the group C BRICHOS containing protein can form a β-hairpin motif, and the hydrophobic nature of this region may indicate that it can insert into membranes and/or is prone to aggregate.

3.4 Bri Protein Family

The Bri family (Bri1, Bri2 and Bri3) includes three integral transmembrane proteins, also called Itm2a-c, with 27 % identical amino acid residues between the human variants. Bri1 is expressed mainly in chondrogenic and osteogenic tissues [73] and Bri3 has been shown to have a strong expression in brain [74], while Bri2 is ubiquitously expressed with a significant expression in the neurons of the hippocampus and cerebellum in humans [6, 75, 76].

What is known about Bri1 is so far very limited, but it has recently been shown that it it has some effect on the development of OT-I thymocytes and that Bri1 is the target gene of the transcription factor GATA-3 in the same cells [77]. A role of Bri1 in cell differentiation during odontogenesis has furthermore been found and its localization suggested that Bri1 might participate in processing and transport of macromolecules in the Golgi network [78].

Bri2 has an unknown function and structure, and is present in neurons, smooth muscle cells, endothelial cells, astrocytes and microglia [79]. The Bri2 protein undergoes proteolytic processing during maturation through the secretory pathway by several proteases (Fig. 5). Furin or furin-like proteases have been proposed to be responsible for generating the 243-residues membrane bound mature Bri2 (mBri2) and a soluble 23-residue peptide (Bri23) [80–82]. Bri23 contains two Cys residues and two stretches with high β-sheet propensities interrupted by a short

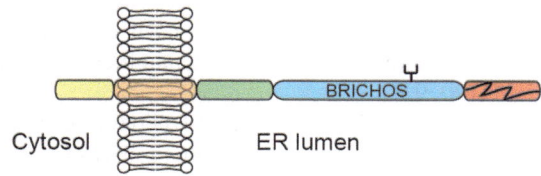

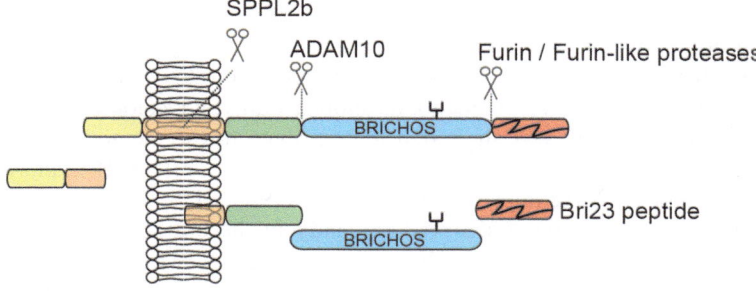

Fig. 5 BRICHOS domain in Bri2. The C-terminal part of Bri2 harbors the Bri23 peptide that is released by furin or furin-like protease. The BRICHOS domain is shedded from the membrane bound Bri2 by ADAM10 protease, and an intra-membrane cleavage by SPPL2 releases the intra-cellular domain of Bri2. An antenna indicates the glycosylation site in the BRICHOS domain. The aggregation prone region in the C-terminal is marked with *wavy black lines*

predicted coil region, suggesting that it can form a β-hairpin structure stabilized by a disulphide bridge [2]. The Bri23 peptide has been proposed to interact with Aβ42 and prevent its aggregation [83], and the Bri2 BRICHOS domain can bind the Bri23 according to mass spectrometry analysis [57]. Bri2 includes one glycosylation site at Asn170, and N-glycosylation has been shown to be important for Bri2 localization at the plasma membrane [84]. The mBri2 can be further processed in the TM region by a signal peptidase like enzyme (SPPL2), and shedding of the extracellular domain including the BRICHOS domain by ADAM10 suggests a physiological extracellular role for the BRICHOS domain [85]. It has also been shown that Bri2 can form a homodimer that is linked by an intermolecular disulfide bridge, but non-covalent interactions are sufficient to keep the homodimer together [86]. Two different mutations in the Bri2 gene results in 11-residue extensions of the C-terminal region, which give rise to peptides, ABri and ADan, that form amyloid fibrils in familial British (FBD) and Danish (FDD) dementia, respectively [87]. ABri and ADan peptides are highly aggregation prone and have been shown to be neurotoxic [88]. FBD and FDD are early onset diseases with gradual progressive dementia and ataxia, and with neuropathological features that include amyloid angiopathy, perivascular plaques, parenchymal deposits of ABri or ADan, and neurofibrillary tangles (NFT) [79]. Bri2 has been shown to suppress

Aβ deposition, suggesting that the levels and/or activity of Bri2 can play a role in AD [83, 89–91]. Bri2 is believed to be a physiological inhibitor of APP process-ing, probably by masking the secretase cleavage sites, and it has been proposed that the loss of wildtype Bri2 affects the levels of APP metabolites, causing similar pathobiology in FDD and AD [90, 91].

Different transgenic (tg) mouse models of FBD and FDD have been developed, but no model has so far fully been able to mimic the human diseases [89, 92–97]. Overexpression of mutant Bri2 in FDD tg mice, gave rise to amyloid deposits sur-rounded by dystrophic neurites, which contained mature wild-type Bri2, and when crossed with a Tau transgene, the mice developed increased amount of tangles, compared with the mice that were transgenic for Tau only [92]. The tg-FDD model gave rise to deposition of ADan amyloid plaques at an early age, while the mice developed behavioural deficits first at a much older age [89]. Knock-in models of FDD and FBD, on the contrary did not develop any amyloid lesions, but decreased levels of wild-type Bri2 was suggested to contribute to the memory deficits and impaired synaptic plasticity described, pointing towards a role of Bri2 in synaptic dysfunction and cognitive impairment in dementia [95, 96].

The Bri3 protein is expressed predominantly in the brain, but has also been found in plasmacytoid dendritic cells, and granulocytes [74, 80, 98]. Bri3 was found to co-localize with APP in a neuroblastoma cell line, and to specifically interfere with the processing of APP, leading to a reduction of Aβ levels by Bri3 overexpression [99].

Recent reports have further pointed towards the importance of Bri2 and Bri3 proteins in association with AD. Bri2 was among the twenty most important dis-ease mediators in ApoE4 carriers and late-onset AD cases [100]. Moreover, increased Bri2 levels and deposition in AD hippocampus have recently been found, which was suggested to be due to aberrant processing of Bri2 [101].

4 Aβ Amyloid Formation in Alzheimer's Disease

AD is the most common form of dementia [102] and the two main hallmarks of its pathobiology are misfolding of amyloid-β peptide (Aβ) and tau, leading to amyloid plaques and intracellular neurofibrillary tangles, respectively. Amyloid(-β) precursor protein (APP) can be proteolytically processed by different secretases, in either an amyloidogenic or a non-amyloidogenic pathway, depending on which secretases that cleave the protein. Aβ is generated from its precursor, the 695- to 770-residue APP, by the sequential activities of β-secretase and γ-secretase (reviewed in [103]). APP is suggested to be involved in neurite growth, plasticity and cell adhesion, but the physiological role for Aβ is unknown [104]. β-secretase was identified at the end of the last century and dubbed beta-site APP cleaving enzyme (BACE) [105–108]. This enzyme cleaves APP on the ER lumenal side, generating soluble APP and a 99-resi-due membrane-bound fragment called C99. The nature of the γ-secratase has been more elusive, but it was shown that four components are necessary and sufficient:

presenilin, nicastrin, Aph-1 and Pen-2 [109, 110]. This complex cleaves not only APP, but also other type I membrane proteins such as Notch [111]. Presenilin seems to be the component that mediates the actual cleavage generating Aβ40 and Aβ42, critically depending on two aspartyl residues [112, 113].

The amyloidogenic pathway generates Aβ peptides of different lengths (38–43 residues), most commonly Aβ40 and Aβ42, whereof Aβ42 is the variant that has shown to be most aggregation prone and toxic. The released Aβ peptide will leave the hydrophobic environment of a phospholipid bilayer, making the discordant Aβ peptide (containing an α-helix composed of β-sheet prone residues, a feature initially discovered in SP-C, Aβ and the prion protein [7]), prone to misfold into β-strand conformation, and aggregation can take off [114, 115]. Many studies indicate that pre-fibrillar intermediates present during the aggregation process of Aβ trigger neuronal dysfunction, rather than the mature fibrils per se [115]. Strategies for AD treatment have so far focused on decreasing the amount of aggregating Aβ by blocking the secretase enzymes, but also attempts to reduce tau protein levels, inhibiting Aβ misfolding, or eliminating the toxic forms of Aβ by immunotherapies have been tried. Unfortunately the results have been disappointing and no disease-modifying treatment is yet available [116].

4.1 Aβ Fibril Formation

Amyloid fibril formation of Aβ is a nucleated polymerization reaction that includes two kinds of nucleation events, primary and secondary nucleation (see [117–120]). The aggregation can be studied using Thioflavin T (ThT), which exhibits enhanced fluorescence upon binding to amyloid fibrils, and a sigmoidal-like growth curve appear when aggregation is monitored as a function of time. The kinetics of Aβ aggregation starts with a lag phase, followed by a rapid process where new fibrils are formed, and at equilibrium a plateau phase is reached [119]. In primary nucleation, monomers associate and form a nucleus, from which a fibril can start to elongate. During secondary nucleation, monomers attach in close proximity on the surface of a fibril, which catalyses the formation of a new nucleus and leads to the exponential fibril growth. This autocatalytic feedback loop has been shown to be the main source for generating toxic oligomers, making the secondary nucleation an attractive target for decreasing the formation of toxic Aβ species (Fig. 6) [117].

4.2 BRICHOS Delays Aβ Aggregation and prevents Toxicity in Vitro

The effect of recombinant BRICHOS domains on Aβ aggregation has been studied using ThT fluorescence as a fibril reporter, and this showed that sub-stoichiometric amounts of recombinant BRICHOS domains of human proSP-C and Bri2 inhibit fibril formation of both Aβ40 and Aβ42 in vitro. A molar ratio of only 0.01 of

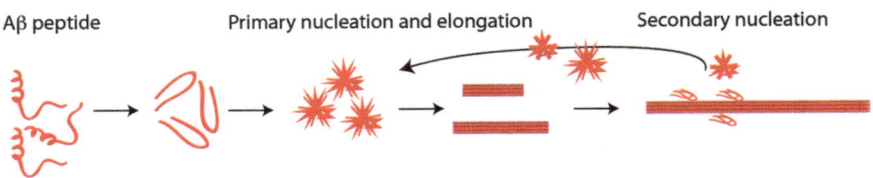

Fig. 6 Aβ fibril formation. The Aβ peptide consists of a discordant helix that can convert into a β-hairpin, which aggregates into oligomers building up a nucleus (primary nucleation) from which fibrils can be formed and elongated. During secondary nucleation, monomers attach in close proximity on the surface of a fibril, which catalyses the formation of a new nucleus and leads to exponential fibril growth [117]

proSP-C BRICHOS in relation to Aβ40 monomer concentration is required for doubling of the aggregation lag time, and with similar amounts of Bri2 BRICHOS there is a 10-fold increase in lag time. For inhibition of Aβ42 fibril formation, more BRICHOS protein is needed compared with the case for Aβ40, but still a molar ratio of 0.6 BRICHOS compared to Aβ42 efficiently delays the aggregation and doubles the lag time (Fig. 7). ProSP-C and Bri2 BRICHOS are able to keep Aβ in a monomeric unstructured state under the prolonged lagphase, and both are capable of halting an already started fibril formation of Aβ, while they cannot dissolve already formed fibrils [58].

Recently, the BRICHOS domain of Gastrokine-1 was shown to also exhibit anti-amyloidogenic properties towards Aβ, where gastrokine 1 prevented Aβ40 from aggregation as shown by SDS-PAGE and ThT fluorescence [121]. By mass spectroscopy a 1:1 complex between gastrokine 1 BRICHOS and Aβ40 was reported [121], and a corresponding complex has been detected previously between Bri2 BRICHOS and Aβ40 [57].

BRICHOS was concluded to interfere with processes that occur during the lag phase of fibrillation [58], i.e. BRICHOS most likely interferes with nucleation events. When Aβ was allowed to fibrillate in vitro in the presence of BRICHOS, it was shown that BRICHOS binds to the fibril surface of Aβ, when analysed by electron microscopy with immuno-gold labelling [56, 122]. The BRICHOS domains investigated so far have been shown to be potent inhibitors of Aβ aggregation in vitro, according to ThT fluorescence and secondary structure analyses, but they do not completely prevent amyloid formation, instead the formation is delayed. In order to be interesting as a possible treatment strategy for AD, BRICHOS also has to show an effect on the toxicity associated with Aβ aggregation, which indeed appears to be the case (see further below).

By fitting analytical data, based on the known rate constants for the different nucleation events, to the experimental kinetics of proSP-C BRICHOS inhibition of Aβ42 aggregation, we have been able to show that BRICHOS specifically inhibits the secondary nucleation of Aβ aggregation (Fig. 8) [122]. BRICHOS binds to the fibril surface and blocks the sites, where otherwise the catalytic event of secondary nucleation takes place, which leads to a significant decrease in oligomer formation. Preventing the major source of oligomeric species by BRICHOS will stop the

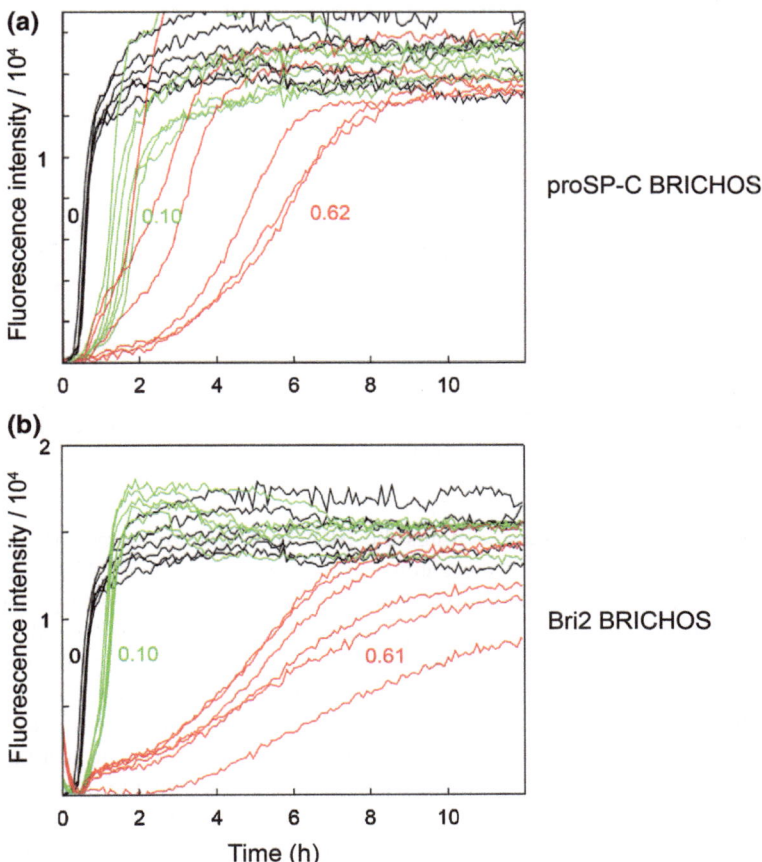

Fig. 7 Aggregation kinetics as monitored by ThT fluorescence. 6 μm Aβ42 at pH 8.0 in the absence (*black*) or presence of 0.10 (*green*, 600 nM) and 0.62 (*red*, 3.7 μM) equivalents of pro-SP-C BRICHOS (**a**); and 6 μM Aβ42 at pH 8.0 in the absence (*black*) or presence of 0.1 (*green*, 600 nM) and 0.61 (*red*, 3.7 μM) equivalents of Bri2 BRICHOS (**b**). This figure was originally published in [58], © the American Society for Biochemistry and Molecular Biology

exponential growth of the fibril, and dramatically reduce the toxicity associated with Aβ aggregation, leaving only the primary nucleation and elongation pathways to build up the fibrils [122].

An important component in the monitoring of brain electrophysiology is recording of so called gamma oscillations, a brain electric activity rhythm that has been suggested to underlie higher cognitive functions [123]. Gamma oscillations are known to be impaired in AD patients [124] and increased Aβ levels has been shown to disrupt the timing of evoked action potentials in a mouse model of AD [125]. By using mouse hippocampal slices, we could recently show that different preparations of Aβ42 peptide reduces the gamma oscillation power in a concentration and time dependent manner and the severity of the reduction is strongest with fibrillated

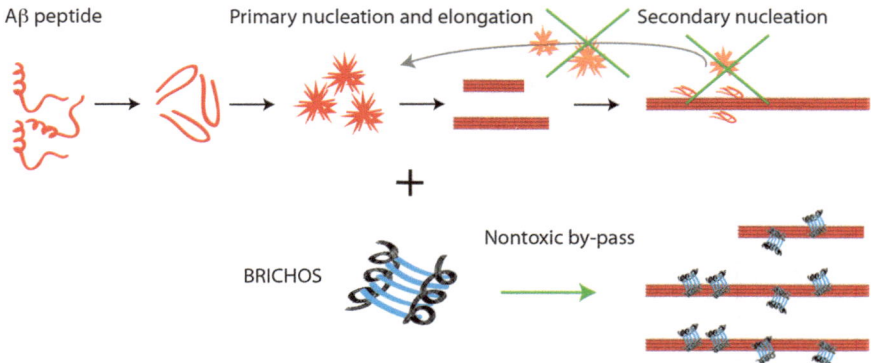

Fig. 8 BRICHOS inhibition of Aβ fibril formation. BRICHOS binds to the surface of the *Aβ* fibril [56], where it specifically protects the sites where secondary nucleation events take place, thereby preventing the catalyzed formation of toxic oligomers [122]

Aβ42 (Fig. 9a) [126]. In these settings the BRICHOS domain of proSP-C is able to prevent the degradation of the gamma-oscillation power by Aβ42, both when the starting Aβ material is a highly pure monomeric preparation or a preparation that also contains mixed species of aggregated Aβ42 (Fig. 9b, c). The BRICHOS domain was also shown to significantly reduce the degradation of gamma-oscillations seen by the fibrillated Aβ42 preparations (Fig. 9c) [126]. More specifically, the BRICHOS domain is able to prevent the Aβ-induced alteration of the excitatory/inhibitory balance in the hippocampal network leading to a protective effect on gamma oscillation power. ProSP-C BRICHOS also prevents the increased toxic effects seen by adding minute amounts of Aβ42 fibrils as seeds to Aβ42 monomers, observed by gamma-oscillation experiments, which also supports the finding that BRICHOS specifically blocks the secondary nucleation [122].

5 In Vivo Effect of BRICHOS on Aβ Aggregation and Toxicity

In 2008 Kim and colleagues reported that expressing a modified version of the Bri2 protein, in which the naturally occurring C-terminal Bri23 peptide was replaced with the Aβ40 peptide (Bri2-Aβ40), in APP transgenic mice by somatic brain transgenic technology, gave rise to less Aβ deposition in the brain [83]. In 2013 this was followed by a study where transgenic mice expressing Bri2-Aβ42 was studied. The mice with Bri2-Aβ42 expression did not only show a slower Aβ deposition and a decrease in oligomeric forms of Aβ42, in addition they did not develop any cognitive impairment despite the fact that they eventually developed plaques [127]. The authors suggested that the delay of Aβ deposition and lack of cognitive impairments point towards possible roles of APP or its

Fig. 9 Prevention of
Aβ induced degradation
of gamma oscillations
by proSP-C BRICHOS.
Power spectra of control
gamma oscillations and
their degradation by 50 nM
monomeric, mixed, and
fibrillar Aβ42 (**a**). Power
spectra showing the
prevention of Aβ induced
degradation of gamma
oscillations by co-incubation
of the proSP-C BRICHOS
domain (**b**). Summary box
plot of prevention of gamma
oscillation degradation by
monomeric, mixed, and
fibrillar Aβ42, in the presence
of BRICHOS (**c**). This figure
was originally published in
[126]

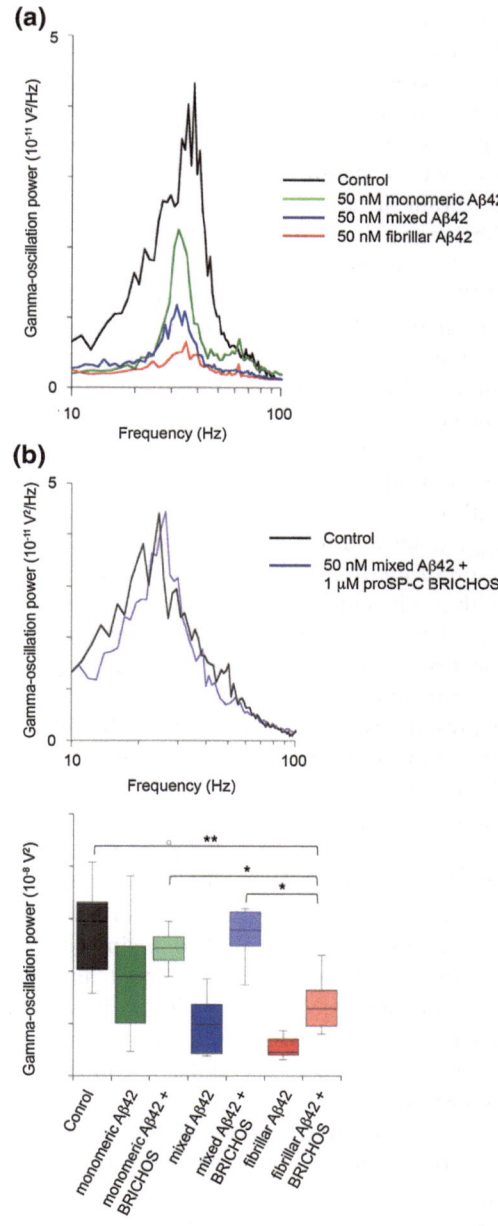

derivatives in Aβ-associated toxicity in mouse models [127]. Another possibility
would be that since overexpressing the Bri2-Aβ42 in the mouse brain results in
that the BRICHOS domain of the Bri2 protein is expressed together with Aβ42,
the lack of cognitive decline can be explained by that the BRICHOS domain

prevents Aβ42 toxicity. This is what the BRICHOS domain has been shown to do in a *Drosophila melanogaster* model of Aβ42 aggregation and toxicity, when it was co-expressed with Aβ42 in neurons. Flies transgenic for the Aβ42 peptide with an upstream signal peptide were crossed with proSP-C BRICHOS Tg flies, and the flies were further crossed with a driver fly line for obtaining flies with a pan-neuronal expression of the transgenes. Expression of Aβ42 in the brain of transgenic *Drosophila* flies has previously been shown to decrease the longevity of the flies and the flies also show impaired locomotor activity, and both events are correlated with Aβ aggregation and deposition in the brain [128, 129]. Crossing Aβ42 transgenic flies with transgenic flies overexpressing the BRICHOS domain from proSP-C, give rise to flies with a life-span similar to that of wild-type flies (Fig. 10a), and the locomotor activity is significantly improved (Fig. 10b) [56]. Analysis of the Aβ protein levels in the Tg flies displayed a difference in the ratio between soluble and in-soluble Aβ, where co-expression of BRICHOS increased the levels of soluble Aβ42 detected in the fly brain at all time-points analysed [56]. The rescue effect on the phenotype seen with co-expression of a BRICHOS domain is evident, even though Aβ42 still aggregates and forms deposits in the brain of the flies, but at a slower rate. In the brain of the fly, co-localization of BRICHOS and Aβ42 was found, and together with in vitro data showing that the BRICHOS domain binds to the surface of an Aβ fibril, this suggest that the interaction between BRICHOS and Aβ prevents the toxicity correlated with the Aβ expression and aggregation [56]. It would be very interesting to analyse if the amyloid deposits and plaques in the mice expressing Bri2-Aβ42, have BRICHOS domain bound to them, in a similar manner as in the fly model. If this is the case, it would support that the BRICHOS domain from Bri2 protein is able to decrease the toxicity of Aβ in vivo also in a mouse model.

Expression of Bri2 without its C-terminal peptide Bri23, i.e. a Bri2 variant in which the BRICHOS domain is located in the very C-terminal end, did not give any effect on Aβ deposition in the APP transgenic mice [83], which would suggest that the BRICHOS domain on its own is not able to the rescue from Aβ toxicity. However, Bri2 without its Bri23 C-terminal peptide was shown to accumulate intracellularly [130]. It is possible that in the case with Bri2 having a very C-terminal BRICHOS domain, it would not be present in the plasma membrane or shedded extracellularly, which could explain the lack of effects on Aβ deposition.

When the FDD mouse model, where the mice overexpress Bri2 with FDD mutation, were crossed with an Aβ-deposition mouse model (APPPS1), the mice showed a significant reduction in the deposition of Aβ in the brain when compared with the single Tg APPPS1. A significant reduction of ADan amyloid deposition was not found in the same mice [89], and even though the BRICHOS domain overexpressed in this model is incapable of preventing the mutant ADan peptide from aggregation, it is possible that it affects and delays Aβ fibril formation.

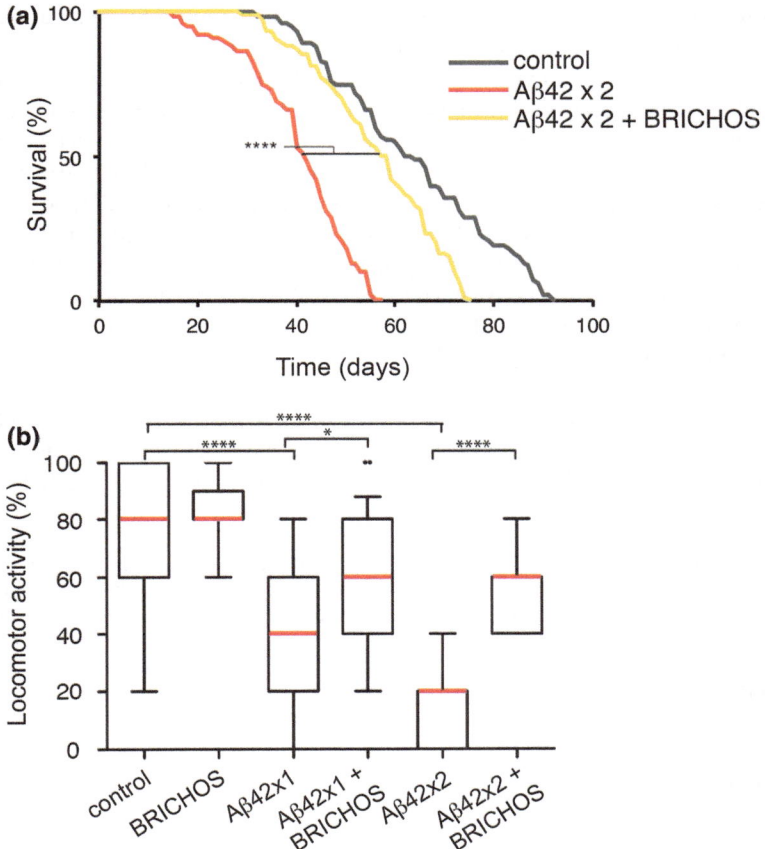

Fig. 10 BRICHOS suppresses the CNS toxic effects of Aβ42 in Drosophila melanogaster. The fraction of 100 living flies over time is plotted for control flies, and flies expressing Aβ42×2 (on two alleles), and Aβ42×2 + BRICHOS. Survival plots were calculated using the Kaplan-Meier method and differences between groups were tested using the log-rank test, ****$P < 0.0001$ (**a**). BRICHOS reduces the toxic effects of Aβ42 on locomotor activity. The number of flies passing an 8 cm line above the ground within 10 s, were counted and expressed as a percentage for each group of flies. Statistical analyses were made using the Mann Whitney test and the *red line* represents the median value. **** = p value < 0.0001, * = p value < 0.05 (**b**). This figure was originally published in [56]

5.1 Future Perspectives

The BRICHOS domains from proSP-C, Bri2, gastrokine-1 [58, 121] and Bri3 (unpublished data) have proven to be very potent inhibitors of amyloid formation, and its presence in a large variety of proteins, makes BRICHOS a ubiquitously expressed natural anti-amyloid chaperone. The recent results showing that BRICHOS also prevents the toxicity associated with Aβ aggregation both in vitro and in vivo, holds promises for targeting this chaperone as a treatment for AD, but also other amyloid diseases. The fibrillation of the aortic-associated amyloid protein medin can also be inhibited by the BRICHOS domain [48], which unmasks the possibility that investigating BRICHOS indeed will be of interest also against other amyloid diseases than AD. Bri2 and Bri3 proteins are expressed in the brain and thus present in the same cell membranes as APP, which makes these two BRICHOS proteins highly interesting for AD treatment. One could think about targeting endogenous BRICHOS by up-regulation of transcription factors, expressing BRICHOS by viral gene transfer or find a way of activating BRICHOS domains already expressed and present in the tissue of interest. Another way would be to use BRICHOS as such as a drug, delivered to the patient. Recombinantly produced BRICHOS has a molecular weight of around 15 kDa, and to get it over the blood brain barrier (BBB), for AD treatment, will be a challenge. However, methods for facilitating passage over the BBB are constantly being developed, which should be further investigated and tested for the BRICHOS protein.

Acknowledgments   We would like to thank all the current and former members working in our group with the BRICHOS domain, as well as our collaborators, who have contributed to increasing the knowledge about the BRICHOS domain and its functions. This work was supported by the Swedish Research Council, The Swedish Alzheimer foundation, The Åke Wibergs foundation, The Magn Bergvalls foundation, Foundation of Gamla tjänarinnor, The Loo and Hans Ostermans foundation for geriatric research.

References

1. Sanchez-Pulido, L., Devos, D., Valencia, A.: BRICHOS: a conserved domain in proteins associated with dementia, respiratory distress and cancer. Trends Biochem. Sci. **27**(7), 329–332 (2002)
2. Hedlund, J., Johansson, J., Persson, B.: BRICHOS—a superfamily of multidomain proteins with diverse functions. BMC Res. Notes **2**(1), 180 (2009)
3. Conkright, J.J., et al.: Secretion of surfactant protein C, an integral membrane protein, requires the N-terminal propeptide. J. Biol. Chem. **276**(18), 14658–14664 (2001)
4. Hayami, T., et al.: Specific loss of chondromodulin-I gene expression in chondrosarcoma and the suppression of tumor angiogenesis and growth by its recombinant protein in vivo. FEBS Lett. **458**(3), 436–440 (1999)
5. Menheniott, T.R., et al.: A novel gastrokine, Gkn3, marks gastric atrophy and shows evidence of adaptive gene loss in humans. Gastroenterology **138**(5), 1823–1835 (2010)
6. Vidal, R., et al.: A stop-codon mutation in the BRI gene associated with familial British dementia. Nature **399**(6738), 776–781 (1999)
7. Kallberg, Y., et al.: Prediction of amyloid fibril-forming proteins. J. Biol. Chem. **276**(16), 12945–12950 (2001)
8. Szyperski, T., et al.: Pulmonary surfactant-associated polypeptide C in a mixed organic solvent transforms from a monomeric alpha-helical state into insoluble beta-sheet aggregates. Protein Sci. **7**(12), 2533–2540 (1998)
9. Willander, H., et al.: High-resolution structure of a BRICHOS domain and its implications for anti-amyloid chaperone activity on lung surfactant protein C. Proc. Natl. Acad. Sci. USA **109**(14), 2325–2329 (2012)

10. Johansson, H., et al.: The Brichos domain-containing C-terminal part of pro-surfactant protein C binds to an unfolded poly-val transmembrane segment. J. Biol. Chem. **281**(30), 21032–21039 (2006)
11. Nerelius, C., et al.: Mutations linked to interstitial lung disease can abrogate anti-amyloid function of prosurfactant protein C. Biochem. J. **416**(2), 201–209 (2008)
12. Robertson, B., Johansson, J., Curstedt, T.: Synthetic surfactants to treat neonatal lung disease. Mol. Med. Today **6**, 119–124 (2000)
13. Weaver, T.E., Whitsett, J.A.: Structure and function of pulmonary surfactant proteins. Semin. Perinatol. **12**(3), 213–220 (1988)
14. Moen, M.D., Perry, C.M., Wellington, K.: Lucinactant: in neonatal respiratory distress syndrome. Treat. Respir. Med. **4**(2), 139–145; discussion 146–147 (2005)
15. Almlen, A., et al.: Surfactant proteins B and C are both necessary for alveolar stability at end expiration in premature rabbits with respiratory distress syndrome. J. Appl. Physiol. (1985) **104**(4), 101–108 (2008)
16. Curstedt, T., et al.: Two hydrophobic low-molecular-mass protein fractions of pulmonary surfactant. Characterization and biophysical activity. Eur. J. Biochem. **168**(2), 255–262 (1987)
17. Curstedt, T., et al.: Low-molecular-mass surfactant protein type 1. The primary structure of a hydrophobic 8-kDa polypeptide with eight half-cystine residues. Eur. J. Biochem. **172**(3), 521–525 (1988)
18. Johansson, J., et al.: Size and structure of the hydrophobic low molecular weight surfactant-associated polypeptide. Biochemistry **27**(10), 3544–3547 (1988)
19. Johansson, J.: Structure and properties of surfactant protein C. Biochim. Biophys. Acta **1408**(2–3), 161–172 (1998)
20. Beers, M.F., Mulugeta, S.: Surfactant protein C biosynthesis and its emerging role in conformational lung disease. Annu. Rev. Physiol. **67**, 663–696 (2005)
21. Stark, M., et al.: Determination of proteins, phosphatidylethanolamine, and phosphatidylserine in organic solvent extracts of tissue material by analysis of phenylthiocarbamyl derivatives. Anal. Biochem. **265**(1), 97–102 (1998)
22. Johansson, J., Szyperski, T., Wuthrich, K.: Pulmonary surfactant-associated polypeptide SP-C in lipid micelles: CD studies of intact SP-C and NMR secondary structure determination of depalmitoyl-SP-C(1-17). FEBS Lett. **362**(3), 261–265 (1995)
23. Curstedt, T., et al.: Hydrophobic surfactant-associated polypeptides: SP-C is a lipopeptide with two palmitoylated cysteine residues, whereas SP-B lacks covalently linked fatty acyl groups. Proc. Natl. Acad. Sci. USA **87**(8), 2985–2989 (1990)
24. Johansson, J., et al.: The NMR structure of the pulmonary surfactant-associated polypeptide SP-C in an apolar solvent contains a valyl-rich alpha-helix. Biochemistry **33**(19), 6015–6023 (1994)
25. Pastrana-Rios, B., et al.: External reflection absorption infrared spectroscopy study of lung surfactant proteins SP-B and SP-C in phospholipid monolayers at the air/water interface. Biophys. J. **69**(6), 2531–2540 (1995)
26. Pastrana, B., Mautone, A.J., Mendelsohn, R.: Fourier transform infrared studies of secondary structure and orientation of pulmonary surfactant SP-C and its effect on the dynamic surface properties of phospholipids. Biochemistry **30**(41), 10058–10064 (1991)
27. Serrano, A.G., Perez-Gil, J.: Protein-lipid interactions and surface activity in the pulmonary surfactant system. Chem. Phys. Lipids **141**(1–2), 105–118 (2006)
28. Gustafsson, M., et al.: Amyloid fibril formation by pulmonary surfactant protein C. FEBS Lett. **464**(3), 138–142 (1999)
29. Carvalheda, C.A., et al.: Structural effects of pH and deacylation on surfactant protein C in an organic solvent mixture: a constant-pH MD study. J. Chem. Inf. Model. **53**(11), 2979–2989 (2013)
30. Gustafsson, M., et al.: The palmitoyl groups of lung surfactant protein C reduce unfolding into a fibrillogenic intermediate. J. Mol. Biol. **310**(4), 937–950 (2001)

31. Baatz, J.E., et al.: Structure and functions of a dimeric form of surfactant protein SP-C: a Fourier transform infrared and surfactometry study. Chem. Phys. Lipids **63**(1–2), 91–104 (1992)

32. Johansson, J., et al.: Secondary structure and biophysical activity of synthetic analogues of the pulmonary surfactant polypeptide SP-C. Biochem. J. **307**(Pt 2), 535–541 (1995)

33. Johansson, J., et al.: Conformational preferences of non-polar amino acid residues: an additional factor in amyloid formation. Biochem. Biophys. Res. Commun. **402**(3), 515–518 (2010)

34. Johansson, J., et al.: A synthetic surfactant based on a poly-Leu SP-C analog and phospholipids: effects on tidal volumes and lung gas volumes in ventilated immature newborn rabbits. J. Appl. Physiol. **95**(5), 2055–2063 (2003)

35. Seehase, M., et al.: New surfactant with SP-B and C analogs gives survival benefit after inactivation in preterm lambs. PLoS ONE **7**(10), e47631 (2012)

36. Sato, A., Ikegami, M.: SP-B and SP-C containing new synthetic surfactant for treatment of extremely immature lamb lung. PLoS ONE **7**(7), e39392 (2012)

37. Glasser, S.W., et al.: Genetic element from human surfactant protein SP-C gene confers bronchiolar-alveolar cell specificity in transgenic mice. Am. J. Physiol. **261**, L349–L356 (1991)

38. Wert, S.E., et al.: Transcriptional elements from the human SP-C gene direct expression in the primordial respiratory epithelium of transgenic mice. Dev. Biol. **156**, 426–443 (1993)

39. Vorbroker, D.K., et al.: Posttranslational processing of surfactant protein C in rat type II cells. Am. J. Physiol. **269**(6 Pt 1), L727–L733 (1995)

40. Beers, M.F., Lomax, C.: Synthesis and processing of hydrophobic surfactant protein C by isolated rat Type II cells. Am. J. Physiol.-Lung Cell. Mol. Physiol. **13**, L744–L753 (1995)

41. Russo, S.J., et al.: Structural requirements for intracellular targeting of SP-C proprotein. Am. J. Physiol. -Lung Cell. Mol. Physiol. **277**(5 Pt 1), L1034–L1044 (1999)

42. Johnson, A.L., et al.: Post-translational processing of surfactant protein-C proprotein: targeting motifs in the NH(2)-terminal flanking domain are cleaved in late compartments. Am. J. Respir. Cell Mol. Biol. **24**(3), 253–263 (2001)

43. Brasch, F., et al.: Involvement of cathepsin H in the processing of the hydrophobic surfactant-associated protein C in type II pneumocytes. Am. J. Respir. Cell Mol. Biol. **26**, 659–670 (2002)

44. Weaver, T.E., Conkright, J.J.: Function of surfactant proteins B and C. Annu. Rev. Physiol. **63**, 555–578 (2001)

45. Mingarro, I., et al.: Different conformations of nascent polypeptides during translocation across the ER membrane. BMC Cell Biology **1**(1), 3 (2000)

46. Johansson, H., et al.: The Brichos domain of prosurfactant protein C can hold and fold a transmembrane segment. Protein Sci. **18**(6), 1175–1182 (2009)

47. Johansson, H., et al.: Preventing amyloid formation by catching unfolded transmembrane segments. J. Mol. Biol. **389**(2), 227–229 (2009)

48. Nerelius, C., et al.: Anti-amyloid activity of the C-terminal domain of proSP-C against amyloid beta-peptide and medin. Biochemistry **48**(17), 3778–3786 (2009)

49. Willander, H., et al.: BRICHOS domain associated with lung fibrosis, dementia and cancer–a chaperone that prevents amyloid fibril formation? FEBS J. **278**(20), 3893–3904 (2011)

50. Glasser, S.W., et al.: Altered stability of pulmonary surfactant in SP-C-deficient mice. Proc. Natl. Acad. Sci. USA **98**(11), 6366–6371 (2001)

51. Clark, J.C., et al.: Targeted disruption of the surfactant protein B gene disrupts surfactant homeostasis, causing respiratory failure in newborn mice. Proc. Natl. Acad. Sci. USA **92**(17), 7794–7798 (1995)

52. Nogee, L.M., et al.: Allelic heterogeneity in hereditary surfactant protein B (SP-B) deficiency. Am. J. Respir. Crit. Care Med. **161**(3 Pt 1), 973–981 (2000)

53. Nogee, L.M., et al.: A mutation in the surfactant protein B gene responsible for fatal neonatal respiratory disease in multiple kindreds. J. Clin. Invest. **93**(4), 1860–1863 (1994)

54. Glasser, S.W., et al.: Pneumonitis and emphysema in sp-C gene targeted mice. J. Biol. Chem. **278**(16), 14291–14298 (2003)
55. Nogee, L.M., et al.: A mutation in the surfactant protein C gene associated with familial interstitial lung disease. N. Engl. J. Med. **344**(8), 573–579 (2001)
56. Hermansson, E., et al.: The chaperone domain BRICHOS prevents CNS toxicity of amyloid-beta peptide in Drosophila melanogaster. Dis. Model. Mech. **7**(6), 659–665 (2014)
57. Peng, S., et al.: The extracellular domain of Bri2 (ITM2B) binds the ABri peptide (1-23) and amyloid beta-peptide (Abeta1-40): Implications for Bri2 effects on processing of amyloid precursor protein and Abeta aggregation. Biochem. Biophys. Res. Commun. **393**(3), 356–361 (2010)
58. Willander, H., et al.: BRICHOS domains efficiently delay fibrillation of amyloid beta-peptide. J. Biol. Chem. **287**(37), 31608–31617 (2012)
59. Casals, C., et al.: C-terminal, endoplasmic reticulum-lumenal domain of prosurfactant protein C—structural features and membrane interactions. FEBS J. **275**(3), 536–547 (2008)
60. Fitzen, M., et al.: Peptide-binding specificity of the prosurfactant protein C Brichos domain analyzed by electrospray ionization mass spectrometry. Rapid Commun. Mass Spectrom. **23**(22), 3591–3598 (2009)
61. Knight, S.D., et al.: The BRICHOS domain, amyloid fibril formation, and their relationship. Biochemistry **52**(43), 7523–7531 (2013)
62. Menheniott, T.R., Kurklu, B., Giraud, A.S.: Gastrokines: stomach-specific proteins with putative homeostatic and tumor suppressor roles. Am. J. Physiol. Gastrointest. Liver Physiol. **304**, G109–G121 (2013)
63. Andra, J., et al.: Structure and mode of action of the antimicrobial peptide arenicin. Biochem. J. **410**(1), 113–122 (2008)
64. Westley, B.R., Griffin, S.M., May, F.E.: Interaction between TFF1, a gastric tumor suppressor trefoil protein, and TFIZ1, a brichos domain-containing protein with homology to SP-C. Biochemistry **44**(22), 7967–7975 (2005)
65. Hnia, K., et al.: Biochemical properties of gastrokine-1 purified from chicken gizzard smooth muscle. PLoS ONE **3**, e3854 (2008)
66. Oien, K.A., et al.: Gastrokine 1 is abundantly and specifically expressed in superficial gastric epithelium, down-regulated in gastric carcinoma, and shows high evolutionary conservation. J Pathol. **203**(3), 789–797 (2004)
67. Oien, K.A., et al.: Profiling, comparison and validation of gene expression in gastric carcinoma and normal stomach. Oncogene **22**(27), 4287–4300 (2003)
68. Dai, J., et al.: Gastrokine-2 is downregulated in gastric cancer and its restoration suppresses gastric tumorigenesis and cancer metastasis. Tumour Biol. **35**(5), 4199–4207 (2014)
69. Otto, W.R., et al.: Identification of blottin: a novel gastric trefoil factor family-2 binding protein. Proteomics **6**(15), 4235–4245 (2006)
70. Azizan, A., Holaday, N., Neame, P.J.: Post-translational processing of bovine chondromodulin-I. J. Biol. Chem. **276**(26), 23632–23638 (2001)
71. Shukunami, C., Oshima, Y., Hiraki, Y.: Molecular cloning of tenomodulin, a novel chondromodulin-I related gene. Biochem. Biophys. Res. Commun. **280**(5), 1323–1327 (2001)
72. Yamana, K., et al.: Molecular cloning and characterization of CHM1L, a novel membrane molecule similar to chondromodulin-I. Biochem. Biophys. Res. Commun. **280**(4), 1101–1106 (2001)
73. Deleersnijder, W., et al.: Isolation of markers for chondro-osteogenic differentiation using cDNA library subtraction. Molecular cloning and characterization of a gene belonging to a novel multigene family of integral membrane proteins. J. Biol. Chem. **271**(32), 19475–19482 (1996)
74. Vidal, R., et al.: Sequence, genomic structure and tissue expression of Human BRI3, a member of the BRI gene family. Gene **266**(1–2), 95–102 (2001)
75. Akiyama, H., et al.: Expression of BRI, the normal precursor of the amyloid protein of familial British dementia, in human brain. Acta Neuropathol. **107**(1), 53–58 (2004)

76. Pittois, K., et al.: Genomic organization and chromosomal localization of the Itm2a gene. Mamm. Genome **10**(1), 54–56 (1999)
77. Tai, T.S., Pai, S.Y., Ho, I.C.: Itm2a, a target gene of GATA-3, plays a minimal role in regulating the development and function of T cells. PLoS ONE **9**(5), e96535 (2014)
78. Kihara, M., et al.: Itm2a expression in the developing mouse first lower molar, and the subcellular localization of itm2a in mouse dental epithelial cells. PLoS ONE **9**(7), e103928 (2014)
79. Rostagno, A., et al.: Chromosome 13 dementias. Cell. Mol. Life Sci. **62**(16), 1814–1825 (2005)
80. Choi, S.C., et al.: Cloning and characterization of a type II integral transmembrane protein gene, Itm2c, that is highly expressed in the mouse brain. Mol. Cells **12**(3), 391–397 (2001)
81. Kim, S.H., et al.: Proteolytic processing of familial British dementia-associated BRI variants: evidence for enhanced intracellular accumulation of amyloidogenic peptides. J. Biol. Chem. **277**(3), 1872–1877 (2002)
82. Kim, S.H., et al.: Furin mediates enhanced production of fibrillogenic ABri peptides in familial British dementia. Nat. Neurosci. **2**(11), 984–988 (1999)
83. Kim, J., et al.: BRI2 (ITM2b) inhibits Abeta deposition in vivo. J. Neurosci. **28**(23), 6030–6036 (2008)
84. Tsachaki, M., et al.: Glycosylation of BRI2 on asparagine 170 is involved in its trafficking to the cell surface but not in its processing by furin or ADAM10. Glycobiology **21**(10), 1382–1388 (2011)
85. Martin, L., et al.: Regulated intramembrane proteolysis of Bri2 (Itm2b) by ADAM10 and SPPL2a/SPPL2b. J. Biol. Chem. **283**(3), 1644–1652 (2008)
86. Tsachaki, M., et al.: BRI2 homodimerizes with the involvement of intermolecular disulfide bonds. Neurobiol. Aging **31**(1), 88–98 (2010)
87. El-Agnaf, O., et al.: Properties of neurotoxic peptides related to the Bri gene. Protein Pept. Lett. **11**(3), 207–212 (2004)
88. Gibson, G., et al.: Structure and neurotoxicity of novel amyloids derived from the BRI gene. Biochem. Soc. Trans. **33**(Pt 5), 1111–1112 (2005)
89. Coomaraswamy, J., et al.: Modeling familial Danish dementia in mice supports the concept of the amyloid hypothesis of Alzheimer's disease. Proc. Natl. Acad. Sci. USA **107**(17), 7969–7974 (2010)
90. Matsuda, S., et al.: BRI2 inhibits amyloid beta-peptide precursor protein processing by interfering with the docking of secretases to the substrate. J. Neurosci. **28**(35), 8668–8676 (2008)
91. Tamayev, R., et al.: APP heterozygosity averts memory deficit in knockin mice expressing the Danish dementia BRI2 mutant. EMBO J. **30**(12), 2501–2509 (2011)
92. Garringer, H.J., et al.: Increased tau phosphorylation and tau truncation, and decreased synaptophysin levels in mutant BRI2/tau transgenic mice. PLoS ONE **8**(2), e56426 (2013)
93. Giliberto, L., et al.: Generation and initial characterization of FDD knock in mice. PLoS ONE **4**(11), e7900 (2009)
94. Pickford, F., et al.: Modeling familial British dementia in transgenic mice. Brain Pathol. **16**(1), 80–85 (2006)
95. Tamayev, R., et al.: Memory deficits due to familial British dementia BRI2 mutation are caused by loss of BRI2 function rather than amyloidosis. J. Neurosci. **30**(44), 14915–14924 (2010)
96. Tamayev, R., et al.: Danish dementia mice suggest that loss of function and not the amyloid cascade causes synaptic plasticity and memory deficits. Proc. Natl. Acad. Sci. USA **107**(48), 20822–20827 (2010)
97. Vidal, R., et al.: Cerebral amyloid angiopathy and parenchymal amyloid deposition in transgenic mice expressing the Danish mutant form of human BRI2. Brain Pathol. **19**(1), 58–68 (2009)
98. Rissoan, M.C., et al.: Subtractive hybridization reveals the expression of immunoglobulin-like transcript 7, Eph-B1, granzyme B, and 3 novel transcripts in human plasmacytoid dendritic cells. Blood **100**(9), 3295–3303 (2002)

99. Matsuda, S., Matsuda, Y., D'Adamio, L.: BRI3 inhibits amyloid precursor protein processing in a mechanistically distinct manner from its homologue dementia gene BRI2. J. Biol. Chem. **284**(23), 15815–15825 (2009)
100. Rhinn, H., et al.: Integrative genomics identifies APOE epsilon4 effectors in Alzheimer's disease. Nature **500**(7460), 45–50 (2013)
101. Del Campo, M., et al.: BRI2-BRICHOS is increased in human amyloid plaques in early stages of Alzheimer's disease. Neurobiol. Aging **35**, 1596–1604 (2014)
102. Fratiglioni, L., et al.: Incidence of dementia and major subtypes in Europe: A collaborative study of population-based cohorts. Neurologic Diseases in the Elderly Research Group. Neurology **54**(11 Suppl 5), S10-5 (2000)
103. Evin, G., Weidemann, A.: Biogenesis and metabolism of Alzheimer's disease Ab amyloid peptides. Peptides **23**(7), 1285–1297 (2002)
104. Dodart, J.C., Mathis, C., Ungerer, A.: The b-amyloid precursor protein and its derivatives: from biology to learning and memory processes. Rev. Neurosci. **11**(2–3), 75–93 (2000)
105. Vassar, R., et al.: Beta-secretase cleavage of Alzheimer's amyloid precursor protein by the transmembrane aspartic protease BACE. Science **286**(5440), 735–741 (1999)
106. Sinha, S., et al.: Purification and cloning of amyloid precursor protein beta-secretase from human brain. Nature **402**(6761), 537–540 (1999)
107. Yan, R., et al.: Membrane-anchored aspartyl protease with Alzheimer's disease b- secretase activity. Nature **402**(6761), 533–537 (1999)
108. Hussain, I., et al.: Identification of a novel aspartic protease (Asp 2) as b-secretase. Mol. Cell. Neurosci. **14**(6), 419–427 (1999)
109. Edbauer, D., et al.: Reconstitution of g-secretase activity. Nat. Cell Biol. **5**(5), 486–488 (2003)
110. Kimberly, W.T., et al.: Gamma-secretase is a membrane protein complex comprised of presenilin, nicastrin, Aph-1, and Pen-2. Proc. Natl. Acad. Sci. USA **100**(11), 6382–6387 (2003)
111. Kimberly, W.T., et al.: Notch and the amyloid precursor protein are cleaved by similar g-secretase(s). Biochemistry **42**(1), 137–144 (2003)
112. Wolfe, M.S., et al.: Two transmembrane aspartates in presenilin-1 required for presenilin endoproteolysis and g-secretase activity. Nature **398**(6727), 513–517 (1999)
113. Kimberly, W.T., et al.: The transmembrane aspartates in presenilin 1 and 2 are obligatory for g-secretase activity and amyloid b-protein generation. J. Biol. Chem. **275**(5), 3173–3178 (2000)
114. Haass, C.: Take five–BACE and the gamma-secretase quartet conduct Alzheimer's amyloid beta-peptide generation. EMBO J. **23**(3), 483–488 (2004)
115. Hardy, J., Selkoe, D.J.: The amyloid hypothesis of Alzheimer's disease: progress and problems on the road to therapeutics. Science **297**(5580), 353–356 (2002)
116. Mangialasche, F., et al.: Alzheimer's disease: clinical trials and drug development. Lancet Neurol. **9**(7), 702–716 (2010)
117. Cohen, S.I., et al.: Proliferation of amyloid-beta42 aggregates occurs through a secondary nucleation mechanism. Proc. Natl. Acad. Sci. USA **110**(24), 9758–9763 (2013)
118. Finder, V.H., Glockshuber, R.: Amyloid-beta aggregation. Neurodegener Dis. **4**(1), 13–27 (2007)
119. Hellstrand, E., et al.: Amyloid ß-protein aggregation produces highly reproducible kinetic data and occurs by a two-phase process. ACS Chem. Neurosci. **1**(1), 13–18 (2010)
120. Jarrett, J.T., Lansbury Jr, P.T.: Amyloid fibril formation requires a chemically discriminating nucleation event: studies of an amyloidogenic sequence from the bacterial protein OsmB. Biochemistry **31**(49), 12345–12352 (1992)
121. Altieri, F., et al.: Anti-amyloidogenic property of human gastrokine 1. Biochimie **106**, 91–100 (2014)
122. Cohen, S.I.A., et al.: The molecular chaperone Brichos breaks the catalytic cycle that generates toxic Aβ oligomers. Nat. Struct. Mol. Biol. (2015). doi:10.1038/nsmb.2971
123. Singer, W.: Synchronization of cortical activity and its putative role in information processing and learning. Annu. Rev. Physiol. **55**, 349–374 (1993)

124. Uhlhaas, P.J., Singer, W.: Neural synchrony in brain disorders: relevance for cognitive dysfunctions and pathophysiology. Neuron **52**(1), 155–168 (2006)
125. Stern, E.A., et al.: Cortical synaptic integration in vivo is disrupted by amyloid-beta plaques. J. Neurosci. **24**(19), 4535–4540 (2004)
126. Kurudenkandy, F.R., et al.: Amyloid-beta-induced action potential desynchronization and degradation of hippocampal gamma oscillations is prevented by interference with peptide conformation change and aggregation. J. Neurosci. **34**(34), 11416–11425 (2014)
127. Kim, J., et al.: Normal cognition in transgenic BRI2-Abeta mice. Mol. Neurodegener **8**, 15 (2013)
128. Crowther, D.C., et al.: Intraneuronal Abeta, non-amyloid aggregates and neurodegeneration in a Drosophila model of Alzheimer's disease. Neuroscience **132**(1), 123–135 (2005)
129. Greeve, I., et al.: Age-dependent neurodegeneration and Alzheimer-amyloid plaque formation in transgenic Drosophila. J. Neurosci. **24**(16), 3899–3906 (2004)
130. Matsuda, S., et al.: Maturation of BRI2 generates a specific inhibitor that reduces APP processing at the plasma membrane and in endocytic vesicles. Neurobiol. Aging **32**(8), 1400–1408 (2011)